# All Kinds of Rocks

written by Molly Bridger
illustrated by Michael Chesworth

HARCOURT BRACE & COMPANY
Orlando Atlanta Austin Boston San Francisco Chicago Dallas New York
Toronto London

Alec collects rocks.

Big rocks.
Small rocks.

Smooth rocks.
Rough rocks.

Alec never has enough rocks.

Some are purple.
Some are white.
Some look dull.
Some look bright.

Some are green and some
are gray.
Some are hard or as soft
as clay.

Some are flat and
some are round.
Alec finds rocks underground.

Alec finds rocks by the sea.
That is where he found
these three.

Alec has some favorite rocks.
He keeps them in
a wooden box.

He likes to show them
to his class.
This one is flat and looks
like glass.

This one is white and has a ring.

This one writes on anything.

This one's beautiful—black and gold!

This one just feels nice to hold.

Alec keeps looking.
He's really tough.
When it comes to rocks,
he can't get enough.